INSIDE THE CHIP

Helen Davies and Mike Wharton

Electronics consultant: Peter Dartnell of STC Semiconductors

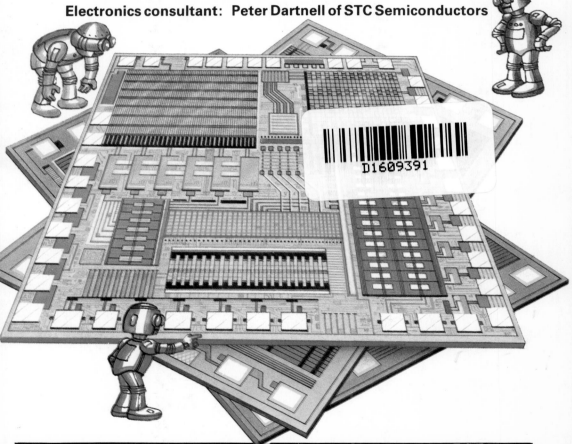

Designed by Graham Round, Iain Ashman and Roger Priddy
Illustrated by Graham Round, Mark Longworth, Graham Smith, Martin Newton, Jeremy Banks, Jeremy Gower, Chris Lyon and Jim Bamber

Contents

- 3 About this book
- 4 Introducing the chip
- 6 How the chip took over
- 8 More about electronics
- 10 How chips work
- 12 Designing a chip
- 14 How chips are made
- 16 Types of chips
- 18 Memory chips
- 20 How memory chips work
- 22 A microprocessor chip
- 24 The registers
- 26 The control circuits
- 28 The microprocessor's clock
- 30 How information gets in and out
- 32 More about input and output
- 34 Inside the ALU
- 36 A logic circuit to build
- 40 How the ALU does arithmetic
- 42 Story of the chip
- 44 Hints on building circuits
- 46 Microprocessor pin chart
- 47 Chip words
- 48 Index

About this book

The chip, more properly called a silicon chip or integrated circuit, is probably the most important invention of this century. To most people the chip is mystifying because it is so small and seems to be able to do so much

This book is a journey inside the chip. It explains what a chip is and how it works, and describes some of the amazing things chips can do. You can find out about the different types of chips and what they are for, and the meaning of the many jargon words which have been invented to describe them.

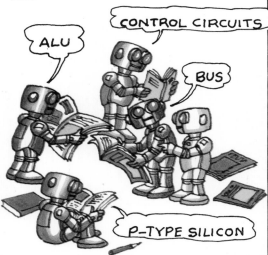

If you looked under the keyboard of a microcomputer, or inside a washing machine or TV or calculator you would see the small box which is the case for the chip. You could also find chips in cars, watches, telephones, video recorders, hospital equipment, traffic lights, robots and factories. This book explains how a chip can control all these things and there is an electronic circuit to build to help you understand how a chip works.

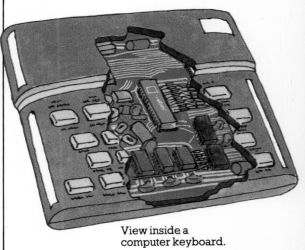

View inside a computer keyboard.

Since its invention in 1958 the chip has developed incredibly quickly and is continuing to develop. Each new robot or more powerful computer is the result of new and more powerful chips. One of the latest developments, still being worked on, is a chip which can recognize and reproduce the immense complexity and variety of human speech. This and all the chip's achievements have made use of just a few basic electronic principles which are explained in this book. You can find out, too, about the scientific discoveries which led to the invention of the chip.

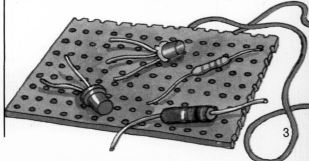

Introducing the chip

A chip is a tiny flake of a substance called silicon, about half the size of a finger nail. It is covered with microscopic electric circuits through which pulse millions of tiny currents of electricity. Using these currents a chip can perform all the operations necessary to control computers, robots, spacecraft, calculators and all sorts of other equipment.

Using electric currents to perform operations is called electronics. A chip is said to be microelectronic because its circuits and the currents it uses are so small.

This picture shows the actual size of a chip (about 5mm square). You can see how small it is in comparison with the sweets.

How a chip controls equipment

Using electric currents as signals, a chip can send and receive messages, do calculations, compare information and make simple logical decisions like a minute electronic brain. This is called processing information and another name for a chip is a microprocessor. Your brain is processing information all the time, as you can see here.

1 To make a bike move, your brain sends out signals to different parts of your body.

2 If something gets in the way your eyes send a message to your brain and your brain processes it.

3 As a result of its work your brain sends out a new set of signals to cope with the new situation.

A microprocessor enlarged many times so you can see the intricate patterns of electric circuits.

A chip can carry out similar functions to a brain but it cannot think in the way humans do. A chip needs a set of instructions, called a program, for everything it does. If the dog in the pictures above were a cat, a microprocessor would run straight into it unless it had been programmed to avoid cats as well as dogs.

Pads

What is electronics?

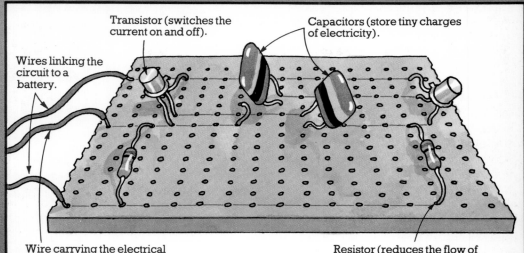

Transistor (switches the current on and off).
Capacitors (store tiny charges of electricity).
Wires linking the circuit to a battery.
Wire carrying the electrical signal produced by the circuit.
Resistor (reduces the flow of the current).

Electronics is the control and manipulation of small electric currents. The devices which control the currents are called components and a circuit is a group of components linked together by wires.

The picture above shows a circuit containing three types of components: transistors, resistors and capacitors. They are linked together by copper tracks which run along the back of the circuit board. The dotted lines show the path which the electric current takes along the tracks between the components.

A microelectronic chip contains many hundreds of circuits, all packed onto one tiny piece of silicon. The components are built *in* and *of* the silicon. They are linked by invisibly fine aluminium tracks which are etched onto the surface of the chip. The proper name for a chip is an integrated circuit, or IC for short. The number of circuits an IC can hold is increasing all the time as better production techniques are developed to fit more components and tracks in the same space.

Packaging chips

Chips are packaged in small plastic cases with legs called pins. The pins are made of copper, coated with gold or tin which are good electrical conductors. They carry electrical signals to and from the chip and also the current needed to power the chip.

If you took the top off the case you would see the chip itself in the centre, with metal connections linking it to the pins. The pattern of metal connections is called the frame or sometimes the "spider". Gold wires, finer than human hair, carry the electrical signals from the spider to the chip. They are welded onto the chip at points round the edge called pads. You can see the pads in the picture on the opposite page.

Pins
Chip
Frame or spider

How the chip took over

The chip was developed during the "space race" of the 1950s and '60s when the Americans were trying to fit more and more electronic equipment into smaller areas. Manufacturers soon exploited the chip's minute size and low power consumption to make pocket calculators and microcomputers. At the same time, because of their reliability and cheapness chips took over from the old mechanical ways of controlling equipment such as watches and cameras, and replaced bulky electronic circuits in TVs, radios and telephone exchanges.

A chip can control almost any machine provided its electronic signals can be translated into a form the machine can use, and information from the machine translated into electronic signals.

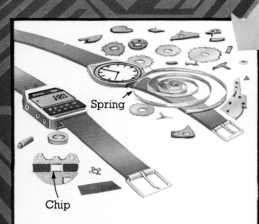

These two watches show the difference between mechanical and microelectronic control. The mechanical watch has many more parts inside, but it can do much less. To power the watch you wind up the spring and as it unwinds it moves the wheels and cogs which, in turn, push the hands round the face. In the electronic watch, the chip measures the time and lights up the figures in the display. It may also display the date and day of the week, and operate an alarm, a timer and even a calculator built into the watch face.

The chip in an electronic watch is powered by small batteries which last for up to five years. It is much more reliable than the cogs and wheels of a mechanical watch because there are no moving parts to break or wear out.

In computers, calculators and video games a keyboard or joystick is used to translate messages from the user into electrical signals which are processed by the chip or chips inside. The results of their work are displayed by using electrical signals to light up different areas of a screen to make numbers, or words and pictures.

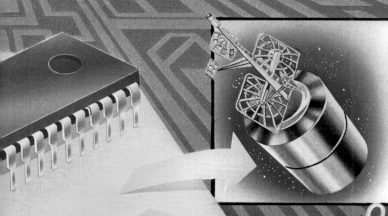

The small size of chips and the incredible speed with which they work are vital in spacecraft where complex calculations must be done in seconds to regulate controls and keep the craft on course. Humans would be incapable of doing these calculations with the speed and accuracy required.

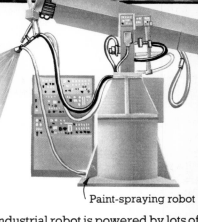

Paint-spraying robot

An industrial robot is powered by lots of separate motors, each of which makes one part of the robot move. By switching different motors on and then off, a chip can make the robot carry out complex sequences of actions.

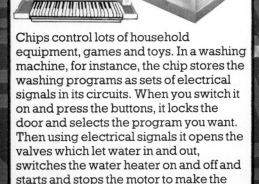

Chips control lots of household equipment, games and toys. In a washing machine, for instance, the chip stores the washing programs as sets of electrical signals in its circuits. When you switch it on and press the buttons, it locks the door and selects the program you want. Then using electrical signals it opens the valves which let water in and out, switches the water heater on and off and starts and stops the motor to make the drum rotate.

This coin-operated telephone hás a chip inside which records electronically how much money you put in, times the call and calculates the cost. If you have any change the chip sends a signal to release the right coins into a dispenser. It will automatically put a call through to the repair engineers if it develops a fault . . . and to the police if it is attacked.

More about electronics

The main components in a chip's circuits are transistors, and it is these which actually do the work. A transistor switches an electric current on and off. It is like an electric light switch, but the switching is done by an electrical force called a voltage, instead of by hand. Transistors are used in lots of electronic equipment, for instance, in TVs, and transistor radios, which are named after transistors.

The pictures below show how a transistor can be used to switch the current to an electric lamp on and off. The circuit is a very simple one but the transistor in it works in the same way as the microscopic transistors in a chip.

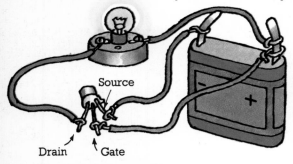 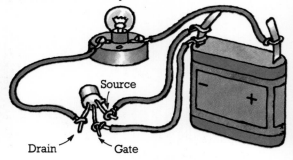

The lamp will not light unless current from the battery can flow round the circuit through the transistor. The three legs of the transistor are called the source, gate and drain. An electric current will only flow through the transistor if an electrical voltage is applied to the gate. For the gate to have a voltage it must be connected to the positive,

or high voltage side, of the battery as shown in the picture on the left. If the gate is connected to the negative or low voltage side of the battery, it will have a low voltage and the lamp will not light. So the lamp can be switched on and off by changing the voltage on the gate.

How transistors work

In the enlarged pictures below you can see inside the metal case of the transistor shown above. The transistor itself is a flake of silicon. Silicon is one of a group of chemicals called semi-conductors which only conduct electricity under certain conditions. To make a transistor, the silicon is injected with impurities which create two new types of silicon with different electrical qualities. They are called p-type and n-type silicon.

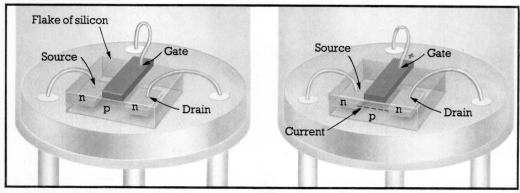

Most transistors have two islands of n-type silicon in a bed of p-type (though they can be made the other way round). The p-type silicon prevents current flowing between the two islands of n-type.

However, when a small voltage is applied to the metal gate above the p-type silicon it changes the p-type silicon near the gate into n-type and the current can flow through the transistor.

Transistors in chips

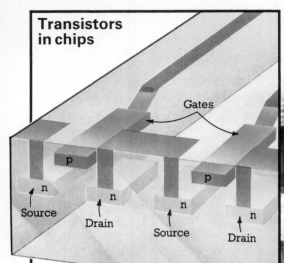

The transistors in a chip are thousands of times smaller than the ones on the opposite page, but they work in the same way. The picture above shows two transistors laid side by side in the surface of a chip. The wires to the source, gate and drain are aluminium tracks laid on top of the transistors and separated from them by an insulating layer of silicon dioxide.

The transistors in a chip's circuits are switched on when they receive a high voltage and off by a low voltage. These voltages are signals which the transistors receive from other transistors as they are switched on and off.

This is what you would see if you looked at a chip through a powerful microscope. The pale green parts are the aluminium tracks going to the source, gate and drain of a single transistor. They are magnified about 2,000 times.

Racing chips

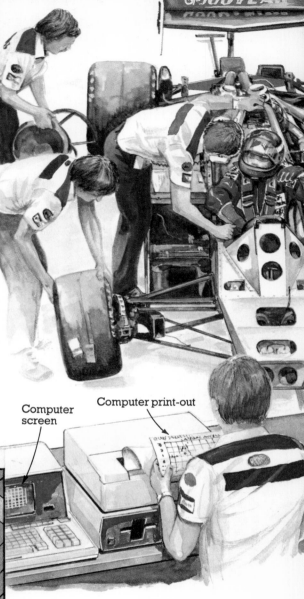

The suspension of this Grand Prix racing car is controlled by chips. The chips are on a circuit board inside the car and during a race they constantly adjust the car's suspension according to the road surface and weight distribution at each moment. Details of the car's performance are transmitted to a computer in the pits by a small aerial behind the driver's head. These show how the suspension system has stood up to different speeds and road conditions, for instance, on corners.

How chips work

As the transistors in a chip's circuits switch on and off they create streams of electrical signals. These are used as a code to represent numbers, letters and any other kind of information. We use codes all the time to represent information and work out problems. Our language and number systems are both codes, and we have another code for musical sounds. A chip has only one code and it is used to represent everything from numbers and words to pictures, sound and movements.

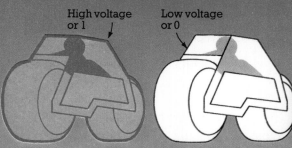

The chip's code has two signals – a high voltage signal, and a low voltage signal. A code with two signals is called a binary code and can be written down in numbers, with a 1 to represent a high voltage and a 0 to represent a low voltage. The 0s and 1s in a chip's code are called bits (short for *bi*nary dig*its*).

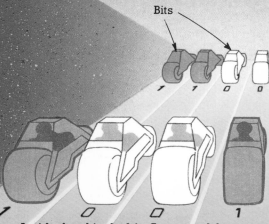

Inside the chip the bits flow round the circuits in groups of eight, called bytes. Each byte is the code for a particular number, letter or other piece of information (e.g. the temperature of a liquid, or the position of a robot arm).

As they travel round the chip the bits move along separate aluminium tracks so they do not get mixed up with one another. A group of eight parallel tracks which can carry one byte at a time is called a bus.

Why is binary used?

When people first tried to make machines which would do calculations they made them using the decimal number system, but the machines were too complex to work well. One of the first was designed by an English mathematician, Charles Babbage, in 1821. He used cogwheels with ten teeth (one for each of the digits 0 to 9) to represent the units, tens, hundreds and thousands columns. The interconnections between all these cogwheels were so complicated that the machine never actually worked.

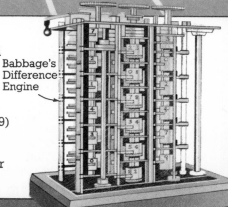

Babbage's Difference Engine

How the code words

Binary is a number code and it works in the same way as our decimal number system. Decimal code has ten digits (0123456789). To make numbers larger than 9 we combine the digits according to a set of rules. Binary uses the same rules for combining its two digits. To understand how binary works think about the rules for counting in decimal.

Decimal

When you count in decimal you go from 0 to 9 and then run out of digits. To make the next number you have to use a combination of two digits. The rules for doing this are to make a new column to the left of the first digit and put a 1 in it, then start again at 0 in the first column.
Each time the first column reaches 9 another 1 is added to the column on its left. When the second column reaches 9 a third column is made on the left of that and so on.

Binary

The rules for counting in binary are exactly the same, but because there are only two digits you can only get as far as 1 before having to start a new column. You start counting 0, 1 as in decimal, but 2 is "10" and 3 is "11". Then for 4 you have to make another column to the left, so it is "100"; 5 is "101", 6 is "110" and so on. Binary seems strange at first but once you have mastered the rules you can do calculations in it, just as you can in decimal.*

Mathematicians had known for centuries that you could count and do calculations using a two digit system, but it was only in 1936 that a German named Zuse had the idea of using binary in a calculating machine. Although binary numbers are too long and unwieldy for humans to work with, they are ideal for machines because they have only two distinct elements instead of ten.
Zuse was an engineering student at the University of Berlin and he lived at home with his parents. He built his first calculating machine, called the Z1, in a corner of the living room and as he added to it and improved it, it gradually filled the whole room.
The Z1 used simple mechanical switches to represent the binary 0s and 1s and light bulbs to show the results of calculations. The switches in Zuse's machine were the forerunners of the transistors on a chip.

*You can find out more about this on page 40.

Designing a chip

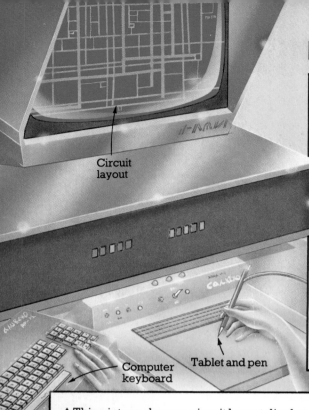

Circuit layout

Computer keyboard

Tablet and pen

A chip is in fact hundreds of separate circuits all linked together and reduced to microscopic size. If the circuits were built using ordinary sized electronic components and wires they would cover the area of a gymnasium.

The job of a chip designer is to devise each individual circuit and then find the best way to link all the circuits together and fit them into the smallest possible space. Originally every circuit had to be drawn by hand from scratch. Now many tried and tested circuits are stored on computer and designers can use these as the basis for a new chip, though they still sketch out new circuits by hand before transferring them to the computer.

▲This picture shows a circuit layout displayed on a computer's TV screen (it is enlarged about 1,000 times). By typing instructions on the computer keyboard, or using a special pad and pen linked to the computer, the designer can add to or alter the circuit, then store the new design back in the computer and call another one onto the screen.

When the circuit designs are complete the designer uses the computer to help work out the best way of linking them all together. The speed with which a chip works depends on how far the electrical signals have to travel between circuits. The time they take is measured in nanoseconds and one nanosecond is one thousand-millionth of a second. The designer's aim is to push the circuits closer and closer together. This not only makes the chip faster, but also more powerful because more circuits can be fitted into the same space.

Check plots

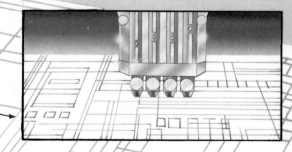

Computer-controlled drawing equipment, called a plotter

When a design is finished the computer holds a list of the exact position of every component and electrical connection. The components are manufactured by building up layers of different chemicals in the surface of the silicon. Maps, called check plots, of each layer of the chip are produced by linking the computer to very precise drawing equipment (shown above). The check plots are about 400 times larger than the chip itself and are used to check the pattern for each layer against the design as a whole.

Testing the design

The production techniques necessary to produce even a single chip are so expensive that test models cannot be made. The final design has to be tested by computer simulation. Simulation means imitating every aspect of a situation so you can see the result of doing something without actually having to do it.

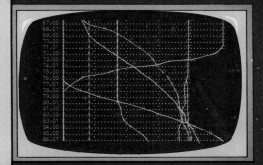 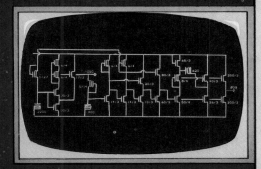

A computer can simulate the way in which electric currents move through a chip's circuits, showing how the transistors will switch on and off and the paths the electrical signals will take through the circuits. When the computer is testing the design it does not actually show the signals moving in the circuits. Instead it displays graphs of voltage changes in particular circuits.

Testing the speed of the new chip is very difficult because it is nearly always faster than any of the chips in the computer which is testing it. This means the computer cannot time the operations of the new chip and complex calculations must be done to estimate its speed.

More about computer simulation

▼Lots of computer games simulate real situations. For instance in computer snooker the computer produces a picture of snooker balls on a table. By pressing control buttons the players can choose the exact position and moment to hit a ball, and the computer then reproduces the distance and direction a real ball would move.

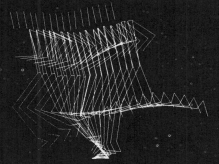

▲Another use for computer simulation is to analyse and improve the performance of athletes. Films of an athlete's movement are converted into a series of stick figures on a computer screen. These recreate the exact movement of each part of the athlete's body. The effect of altering the movement of one part can be investigated by changing it on the computer and studying the effect it has on the stick figures.

How chips are made

To fit many hundreds of separate circuits onto a tiny chip of silicon about 5mm square requires incredibly precise production techniques. The components in a chip's circuits are measured in microns and must be positioned with an accuracy of one or two microns. A micron is a thousandth of a millimetre (this page is about 170,000 microns wide). Chips are made using sophisticated computer-controlled machinery in ultra-clean, dust-free factories. Powerful microscopes are needed to look at the chips during manufacture.

To make a chip the components and circuit connections are built up in layers in and on the surface of the silicon. There may be as many as nine or ten different layers. You can find out how they are made below.

Silicon is a chemical which is part of ordinary sand. To make chips, cylindrical crystals of pure silicon are grown in a vacuum and then sliced into wafers about ½mm thick. The wafers are put into a machine like the one in the picture above, which grinds the surface absolutely smooth. Each wafer will make several hundred chips.

Using the circuit designs held in the computer's memory a set of "photomasks" are made, one for each layer of the chip. The masks are squares of glass on which the pattern of a circuit layer is printed, either by a photographic process or by a more precise technique called electron-beam lithography. The glass masks are about 10cm square and hold the patterns for a few hundred chips side by side.

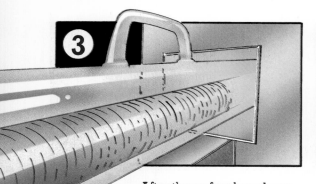

After the wafers have been thoroughly cleaned they are put into a red hot oxidation furnace, where they grow a thin insulating layer of a chemical called silicon dioxide. Then they are coated with a soft, light-sensitive plastic called photoresist. This and the next process are repeated for every circuit layer of the chip.

To transfer the circuit pattern from the mask to the silicon, the mask is positioned over the wafer which is then flooded with ultraviolet light. This hardens the photoresist in the areas not protected by the mask. Acids and solvents are used to strip away the soft, unexposed photoresist and the silicon dioxide beneath it, leaving bare the areas of silicon to be treated.

5 The first layers to be embedded in the silicon are the chemical impurities, called dopants, which produce the n- and p-type parts of the components. The dopants are embedded by a method called ion implantation (ions are electrically charged particles). The wafers are put into a special machine where they are bombarded with ions of the dopant chemical. The ions travel at an enormous speed and hit the wafer with such force that they embed themselves into the exposed areas of the silicon.

Loading wafers into an ion implanter.

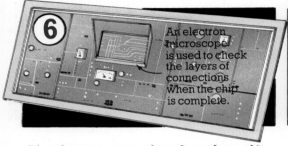

An electron microscope is used to check the layers of connections when the chip is complete.

The aluminium connections are about four microns wide.

After the components have been formed in the silicon, aluminium connections are laid on top. There may be two layers of connections separated from one another by insulating layers of silicon dioxide. The aluminium layers are laid down by an evaporation process, using masks to define the tracks.

This enlarged picture of a tiny area of a chip shows how a single dust particle can cause a break in an aluminium connection and ruin a whole microelectronic circuit. The air in the rooms where chips are made is continuously filtered and recirculated to keep it dust-free, and workers dress from head to foot in overalls.

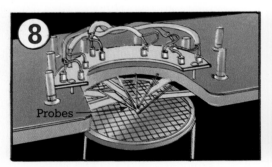

Probes

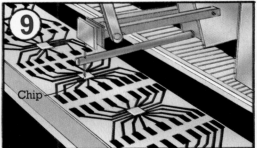

Chip

When the manufacturing is complete electrical probes are used to test each individual chip. As many as 70% are found to be faulty at this stage and marked with a red dot. Then the wafer is sliced into individual chips with a diamond or laser saw and the faulty chips are discarded. The perfect chips are then packaged as shown in the next box.

To package the chips, machines rather like sewing machines weld gold wires to the pads at the edge and attach them to the metal connectors of the frame. Then the plastic casing is put on and the pins bent down. When they are finished the chips are submitted to more tests to check they will continue working in any conditions, e.g. in extreme cold or outer space.

Types of chips

The picture below shows the main types of chips in a home computer. The most important one is the microprocessor. This is the type people usually mean when they talk about "the chip" and it has the circuits necessary to control a computer, or any other machine. However, a microprocessor cannot work on its own. It needs circuits with electrically coded instructions telling it what to do, stores to hold information while it is at work, and coders and decoders to translate electrical signals from the outside world to and from binary code. All these jobs are done by different types of chips.

Microprocessor chip

The microprocessor is the chip which carries out the calculations and logical decisions needed to control a machine. In a computer the microprocessor is often called the Central Processing Unit or CPU chip.

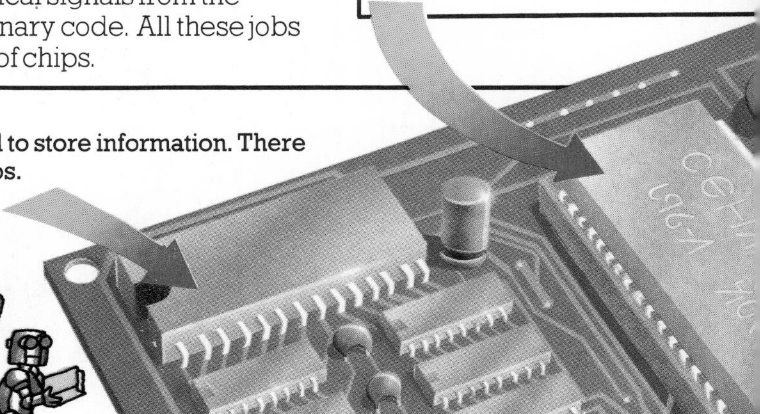

Memory chips

These have circuits designed to store information. There are two types of memory chips.

ROM chip

ROM stands for Read Only Memory. This chip is a permanent store for electrically coded information. The information is stored in the chip when it is made, by setting the transistor switches so they produce the same pattern of signals each time the current flows through them.* You can think of the ROM chip as a library holding sets of instructions, called programs, which tell the microprocessor what to do. The microprocessor can only "read" information in a ROM chip, it cannot store anything new there.

RAM chips

RAM is short for Random Access Memory, though a better name for these chips would be Read/Write Memory. RAM chips are temporary stores where electrically coded information can be "written" or "read" and then rubbed out when it is no longer needed. Each time a new piece of information is stored in a RAM chip, its transistor switches are reset to create the pattern of signals which represents that piece of information.

You can think of RAM chips as a notice board where the microprocessor can keep a temporary record of information it needs for a particular operation.

*A ROM chip made in this way is called a mask programmed ROM. You can find out about other kinds of ROMs on page 47.

Interface chips

These chips change various kinds of electrical signals from the outside world into the binary 0 and 1 signals which the microprocessor and other chips can deal with. They also translate the microprocessor's binary codes back into electrical signals which can be used to light up a computer TV screen or to control machinery.

From outside world

To microprocessor

The clock

Although a microprocessor carries out thousands of operations each second it can only do them one at a time. To keep all the operations in order, a quartz crystal "clock" beats time.

Interconnections

All the chips needed to control a piece of equipment, such as a home computer, are attached to a board called a printed circuit board (PCB for short). This has metal tracks printed on its surface which carry the electrical signals between the microprocessor and the other chips. Some of the tracks lead to an edge connector where other parts of the machine are plugged onto the printed circuit board.

All-in-one chips

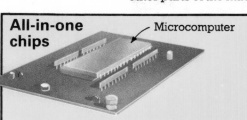

Microcomputer

Most calculators, toys and household equipment such as washing machines are controlled by a single chip which contains all the control, memory and interface circuits in one. This type of chip is called a microcomputer* or a dedicated microprocessor. It is called "dedicated" because the ROM circuits (that is, its instructions) are built onto the chip. This means the microprocessor can only be used for the tasks described in those ROM circuits.

An ordinary microprocessor like the one in the main picture, is not dedicated. Its instructions are on a separate ROM chip and by changing the ROM chip the microprocessor can be made to do a different job. For instance, the same microprocessor could be used to control a computer, or a video game or a robot simply by giving it a different ROM chip.

*Small computers are named after it.

Memory chips

Memory chips are used to store electrically coded instructions and information. Although a microprocessor works incredibly quickly it can carry out only a limited number of simple tasks such as adding two numbers, or comparing two pieces of information. Each task is represented by an eight-bit binary code, called a machine code instruction. To make a microprocessor do any job it must be given a list of machine code instructions telling it each of the tasks involved. These lists of instructions, or programs, are stored in the memory chips.

What a ROM chip holds

A ROM chip holds the programs the microprocessor needs to tell it how to control a machine. These programs are called the Operating System or Monitor. In a computer they tell the microprocessor what to do when the computer is switched on, how to recognize when a key has been pressed, where to store the electrical signals it produces, how to light up the TV screen to make words and pictures, and so on. The ROM chip in a computer also holds a program called the Interpreter which tells the microprocessor how to translate the user's instructions, typed in as words and numbers, into machine code instructions which it can understand.

What RAM chips hold

The RAM chips are temporary stores for information the microprocessor needs for a particular operation. For example, it may store the result of a calculation which it will need again later in a program. In a computer the user's programs of instructions are stored in the RAM chips and the results of the microprocessor's work are also stored there before being displayed on the TV screen. Often a microprocessor copies a program stored permanently in the ROM chip into the RAM chips in order to work through it.

Inside a memory chip

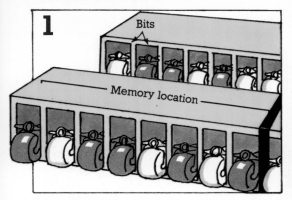

Both ROM and RAM chips can be thought of as containing rows of boxes each of which can hold one bit of data. To a chip, data is either a machine code instruction or coded information. A group of eight boxes which can hold one byte of data is called a memory location. The memory locations are numbered so that the microprocessor can find information and the number of each location is called its address.

Coding addresses

A microprocessor has no way of distinguishing between ROM and RAM chips so the addresses of memory locations in different chips must not overlap. (Imagine every house in a city having to have a different number, instead of repeating numbers with different street names.) The numbers start at 0 in ROM and move up numerically through the RAM chips.

Address numbers are coded into 16-bit binary codes because so many of them are needed. An eight-bit code would only produce 256 different addresses, but sixteen bits can be arranged in 65,535 (2^{16}) different ways and so produce this number of different addresses.

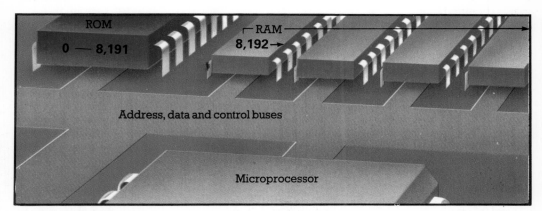

Linking the memories to the microprocessor

In a machine, memory chips are linked to the microprocessor by sets of wires called buses. There are three buses, one for addresses, one for data and a third to carry various control signals. The address bus has 16 wires, each carrying one bit of the address code. The data bus has eight wires because data (i.e. information and instructions) is coded into eight-bit bytes. The control bus is a group of wires which carry signals such as the "Read/Write" signal which tells the memory chip whether data is to be taken from or stored in a particular memory location.

How memory chips work

The pictures below show how information and instructions are sent between the memory chips and the microprocessor.

When the microprocessor wants to look at information in a particular memory location it sends the address code for that location along the address bus to the memory chip. When the code arrives, decoder circuits on the memory chip read the pattern of signals and point to the memory location addressed.

A signal on the control bus indicates whether the memory location is to be read or written into. The memory locations in ROM chips can only be read, but in RAM chips the microprocessor may either read information or write (i.e. store) it.

If the control signal says "read", circuits called sense amplifiers read the code stored in the memory location and put a

copy of it onto the data bus. (The code is duplicated – it does not leave the memory location.)

More about buses

It is easiest to think of buses as groups of parallel wires, but in fact they are tracks printed onto the printed circuit board in whatever way fits best. The buses go right up into the chip itself carried up from the PCB by the pins of the chip's case.

The picture on the right shows a ROM chip with 28 pins. Eight of the pins carry the eight-bit data bus, but there are only 13 pins for the address bus. This is because a single chip does not have 65,536 memory locations, so it does not need all the addresses that can be produced by the 16-bit address bus. In fact, this chip has 8,192 memory locations so it only needs 13 lines to produce all its addresses ($2^{13} = 8,192$).

Pins for data bus
Pins for address bus
Pins for control bus

The remaining pins are for power supply.

What a memory chip looks like

The picture below shows a RAM chip much enlarged so you can see that the circuits are laid out in two blocks. The blocks are made up of thousands of identical circuits, called memory cells, arranged in rows and columns. Each memory cell is formed using about six electronic components and can hold one bit of information. A memory location is eight cells.

The data bus runs between the two blocks of memory cells so that the signal from any particular cell has the shortest possible distance to travel to the data bus. This is important because it affects the chip's "access time". The access time is the time it takes a piece of information to get from the memory to the microprocessor. A typical access time is 200 nanoseconds.

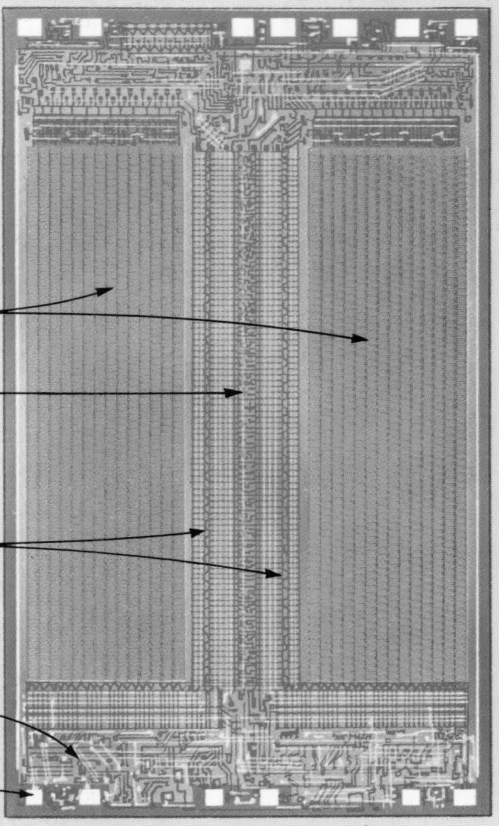

Blocks of memory circuits

Sense amplifiers – circuits which read the data codes stored in the memory locations and put them onto the data bus.

Column decoders – circuits which decide which column a memory location is in. The second byte of the address is the code for the column.

Row decoders – circuits which decide which row a memory location is in. The first byte of the 16-bit address code is the code for the row.

Pads where address, data and control signals enter and leave the chip.

The size of memory chips is usually measured by the number of *bits* (not bytes) they can hold. The chip in this picture can hold 16,384 bits (because there are two blocks each with 128 columns and 64 rows). This number is written as 16k (k stands for kilo which means 1,024 in chip and computer language).*

More about RAM chips

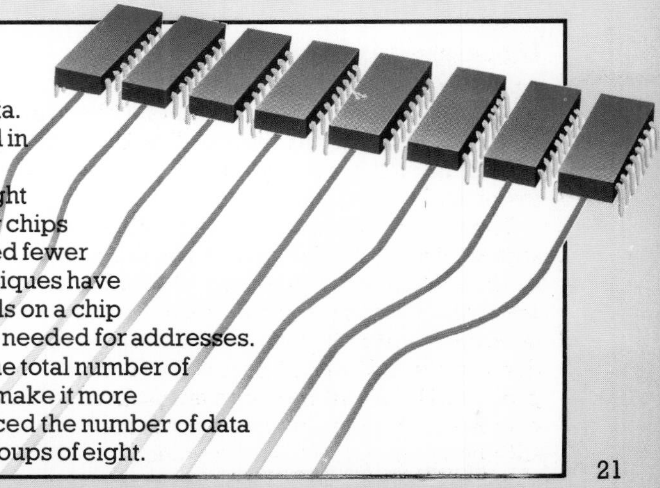

Most RAM chips have only one pin for data. In a machine the RAM chips are arranged in groups of eight and the bytes of data are split up so one bit goes into each of the eight chips. The reason for this is that originally chips had much fewer memory cells and needed fewer addresses. However as production techniques have improved and the number of memory cells on a chip increased, more and more of the pins are needed for addresses. Manufacturers did not want to increase the total number of pins on a chip's case because this would make it more expensive to produce. Instead they reduced the number of data pins to one and always use the chips in groups of eight.

*A capital letter K means a kilo*byte* and the size of memory chips, especially ROMs, is often expressed in kilobytes. For instance, the ROM chip on the opposite page is 8K because it stores 8,192 ($8 \times 1,024$) bytes.

A microprocessor chip

This picture shows the circuits of a microprocessor, the kind of chip that can control a machine such as a robot or computer. A microprocessor controls a machine by sending electrical signals to it and processing, or working on, the information it receives back from the machine. The microprocessor is told what signals to send, how to get information back, how to process it and what to do with the results, by the programs of instructions stored in the memory chips. You can find out more about programs below.

The circuits of a microprocessor can be grouped into three main types: the control circuits, the registers and the ALU circuits. ALU stands for Arithmetic Logic Unit and this is where the main work is done.

Registers

10110100
011011010101 0001

Registers are like memory locations built on to the microprocessor itself. Some have eight boxes and can hold one byte of information, others have 16 and can hold two bytes. The registers are used to store information and instructions inside the microprocessor while they are being worked on, and also to hold addresses which are about to be sent to the memory chips. Some of the registers such as the program counter do very specific jobs. You can find out more about these on the next few pages.

More about programs

Every job a microprocessor does is broken down into a sequence of simple tasks and an electrically coded instruction is needed for each task. On the right you can see an example of the kind of instructions a microprocessor needs, written in English with decimal numbers so you can understand them.

A program is a list of instructions for a particular job and each instruction is coded into an eight-bit byte of binary code. Usually instructions are followed by information, or an address where information can be found, and these are also coded into binary and form part of the program. Addresses always take up two bytes of a program because they are 16 bits long.

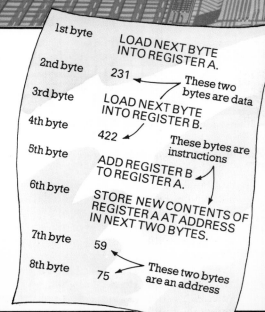

1st byte — LOAD NEXT BYTE INTO REGISTER A.
2nd byte — 231 — These two bytes are data
3rd byte — LOAD NEXT BYTE INTO REGISTER B.
4th byte — 422
5th byte — These bytes are instructions
ADD REGISTER B TO REGISTER A.
6th byte — STORE NEW CONTENTS OF REGISTER A AT ADDRESS IN NEXT TWO BYTES.
7th byte — 59
8th byte — 75 — These two bytes are an address

Arithmetic Logic Unit

This is the part of the microprocessor where the calculations and logical decisions necessary to control a machine are carried out. They are done by manipulating binary codes through circuits which change them in particular ways. In fact electronic circuits can only perform a limited number of simple logic operations. All the calculations and decisions are built up from these. You can find out how this is done on pages 34 to 41.

Control circuits

Nearly two thirds of the surface of a microprocessor is occupied by control circuits. These organize all the microprocessor's work by sending out signals to open and close registers, trigger off operations in the ALU and so on. The control circuits are linked to the microprocessor's clock and you can find out more about how they work on pages 26 to 29.

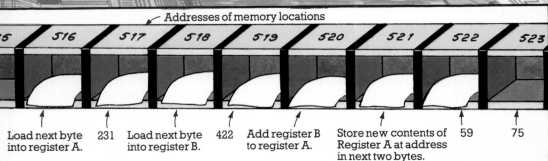

Addresses of memory locations: 515, 516, 517, 518, 519, 520, 521, 522, 523

- Load next byte into register A.
- 231
- Load next byte into register B.
- 422
- Add register B to register A.
- Store new contents of Register A at address in next two bytes.
- 59
- 75

This picture shows how the program on the left would be stored in a memory chip. Each byte of the program is a code of eight electrical signals stored in a memory location. When a microprocessor is at work it fetches the bytes of a program one by one in numerical order. It can only tell whether a particular byte is an instruction, an address or data by its position in the program. For instance the first byte of a program is always an instruction. If a programmer put a byte of data first by mistake the microprocessor would treat the data as an instruction and the whole program would go haywire or "crash".

The registers

About one tenth of the surface of a chip is covered with registers. These are temporary stores for information, instructions and addresses as they are moved round inside the microprocessor. Some of the registers have a special role to play in the operation of the microprocessor and these are described below.

Registers work in the same way as the memory locations on a RAM chip. Data can be read from or written into them, and reading data means taking a copy of it, not actually taking it out of the register. Data is only removed from a register when more data is written over the top of it.

The Program Counter

The Program Counter is a 16-bit register and the binary codes it holds are addresses. Its job is to display the next address from which data or an instruction is to be fetched, so that the program keeps running in the right order.

At the start of a program the Program Counter is loaded with the address of the first instruction (that is, the first byte of the program). This is put onto the address bus.

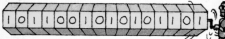

While the instruction is being fetched and carried out the Program Counter is automatically increased by 1, so the number it now holds is the address of the next byte of the program. (It is rather like a counter on a cassette recorder.) The operation which automatically increases the number in the Program Counter by 1 is called incrementing.

The Program Counter will keep displaying addresses in numerical order unless it is told to jump to a different part of the program. When this happens the new address is loaded into the Program Counter and it starts counting again from there.

Instruction Register

Each instruction brought into the microprocessor from the memory chips is sent to this eight-bit register and then carried out by the control circuits. An instruction stays in the Instruction Register until it is replaced by the next instruction brought in from the memory chips.

Flag register

This is an eight-bit register and each of the bits is used separately by the ALU to indicate that a particular event has happened. Each bit in the register is called a flag and signals a different event. If the bit is a 1 (a high voltage) the flag is "set" showing the event has happened. If the bit is a 0 (a low voltage) it shows the event has not happened.

Set the carry flag.

For example one of the flags is called the "carry" flag and it is used by the ALU when it is doing addition. If the result of adding two binary numbers is larger than eight bits (and will not fit into the Accumulator) the ALU indicates this by setting the carry flag.

Accumulator

This is an eight-bit register linked to the ALU. It is used by the ALU when it is doing calculations. For example, for the ALU to add two numbers one of them must be loaded into the Accumulator. When the addition has been done the answer is sent to the Accumulator where it replaces the original number. It is often said that this register "accumulates" results and this is how it gets its name.

Working registers

These are eight-bit registers where data is stored ready for the ALU to work on, or where results are stored temporarily if they will be needed again soon. The number of working registers on a microprocessor varies between six and 30 according to the make. This means that very little data can be held inside the microprocessor at any time. Most of it has to be stored in the RAM chips and brought in each time it is needed.

On some microprocessors the working registers are arranged in pairs so two can be used together to hold address codes.

The control circuits

A microprocessor's control circuits are like the captain of a ship telling everybody what to do and when to do it. Each machine code instruction that arrives in the Instruction Register involves lots of separate little operations, such as opening and closing registers, fetching data from memory, and so on. The control circuits send out signals to make all these operations happen in the right order and at the right time. They are themselves kept in time by the microprocessor's clock which you can find out more about over the page.

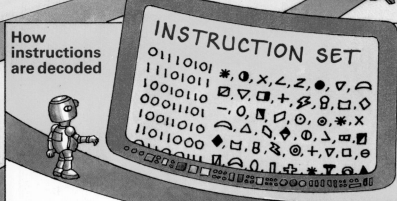

How instructions are decoded

Inside the control circuits there is a list of all the machine code instructions the microprocessor can understand. This list is called the microprocessor's Instruction Set. The number of instructions in it varies between about 70 and 150 depending on the make of microprocessor.

The control circuits contain a small group of memory circuits called the Microprogram ROM. These hold a sequence of signals for each instruction in the Instruction Set. These control signals cause the machine code instructions to be carried out inside the microprocessor.

When a machine code instruction arrives in the Instruction Register its pattern of 0 and 1 signals is used as an address to locate the place in the Microprogram ROM where the control signals for that instruction are stored.

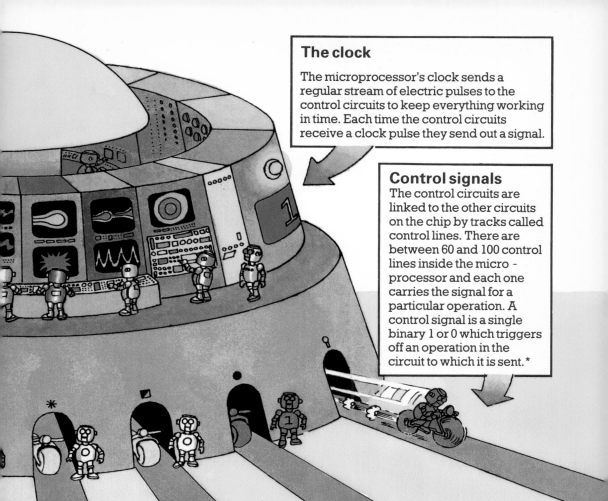

The clock
The microprocessor's clock sends a regular stream of electric pulses to the control circuits to keep everything working in time. Each time the control circuits receive a clock pulse they send out a signal.

Control signals
The control circuits are linked to the other circuits on the chip by tracks called control lines. There are between 60 and 100 control lines inside the micro-processor and each one carries the signal for a particular operation. A control signal is a single binary 1 or 0 which triggers off an operation in the circuit to which it is sent.*

More about control signals

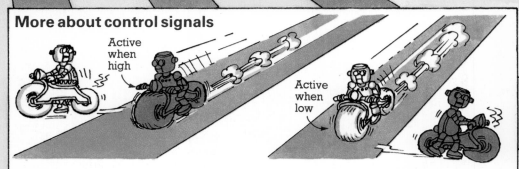

In electronic terms a binary 1 signal is a high voltage and a binary 0 signal is a low voltage. Normally, a control line has a low voltage most of the time, and the signal to do something is given by changing this to a high voltage. (You can think of this as sending a binary 1.) The control line is said to be "active when high".

It can be the other way round though, so the control line is "active when low". This means the line has a high voltage most of the time and to make something happen the high voltage is changed to a low one (this is a binary 0 signal). In fact all the 0 and 1 signals in the chip's code are low and high voltages.

*Some of the control lines form the control bus which carries control signals to chips outside the microprocessor, e.g. to the memory chips.

The microprocessor's clock

Every tiny, electronic operation that happens in a chip's circuits must be triggered by a control signal. The control signals themselves are triggered by the microprocessor's clock.

The clock is a group of circuits linked to a thin slice of quartz crystal. The circuits may be on the microprocessor itself or on a separate clock chip. The quartz crystal is always separate because it is itself as big as a chip.

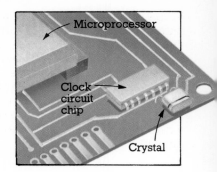

When an electric current is passed through the quartz crystal it vibrates at a very precise and regular rate. The clock circuits use these vibrations to send a regular stream of pulses to the microprocessor's control unit. Each time the control unit receives a pulse it sends out a signal and an operation takes place somewhere on the microprocessor. Nothing can happen without a clock pulse – if the clock missed a pulse every circuit on the chip would be frozen for that instant.

The rate at which the clock sends out pulses is measured in megahertz and it controls the speed at which the microprocessor works. One megahertz is one million pulses per second. A microprocessor's clock may run at between two and four megahertz which means the microprocessor is performing up to four million tiny operations each second.*

More about the clock

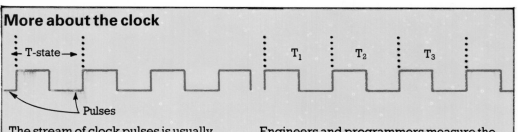

The stream of clock pulses is usually represented by a line like this called a square wave. The part of the line going up represents a pulse and the time interval between that and the next pulse (the next vertical line going up) is called a clock cycle or T-state.

Engineers and programmers measure the time it takes the microprocessor to carry out a machine code instruction in T-states. The number of T-states is the same as the number of electronic operations the instruction requires. It may be anything from four to 20 depending on how complicated the instruction is.

*That is, one every 250 nanoseconds.

How a chip starts working

When a machine controlled by a microprocessor is switched on, the first thing that happens is that an electrical signal is sent direct to the Program Counter to set it to zero. This is called the reset signal and it enters the microprocessor through one of its legs. As soon as the reset signal has been sent, the clock is started and this in turn generates signals from the control circuits.

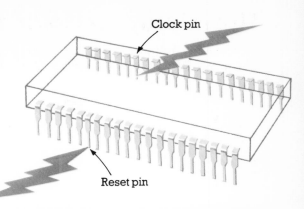

The first control signals put the number in the Program Counter (it is zero, 0000 0000 0000 0000 in binary) onto the address bus and fetch the byte stored at that address into the Instruction Register. The binary number 0000 0000 0000 0000 is always the address of the first memory location in the ROM chip. The byte it holds is the first in a sequence of instructions which prepare the machine for use.

In a computer getting the machine ready involves clearing all the memory locations in the RAM chips, doing a test to check they are storing electrical signals correctly, checking all the input and output addresses to see if equipment is connected to them and finally displaying a message on the screen to let users know they can start typing in programs.

Killing time

All the time the machine is switched on the clock is pulsing, so the microprocessor has to be doing something (even if nothing needs doing at that moment). To prevent the microprocessor racing blindly through program after program special routines called "waiting loops" are written into the controlling programs of the ROM chip. These make the microprocessor work through a series of instructions over and over again until something else comes along for it to do.

Waiting loop

One of the waiting loops in a home computer is a sequence of tests to check whether a key has been pressed on the keyboard. Every time the microprocessor has nothing else to do the controlling programs will direct it to carry out these tests over and over again. In fact it is estimated that a microprocessor in a microcomputer spends 98% of its time looking at the keyboard.

How information gets in and out

A microprocessor is only useful if it can be connected in some way to the outside world, so it can send its signals out and receive information back. A microprocessor controlling a computer or arcade game, for instance, needs to be able to receive the user's instructions and display the results of carrying them out. A microprocessor controlling a machine such as an arm robot or a washing machine must make the machine move and also receive information back about what the machine is doing (this is called feedback). The pieces of equipment which translate "real world" information to and from electrical signals are called input and output devices.

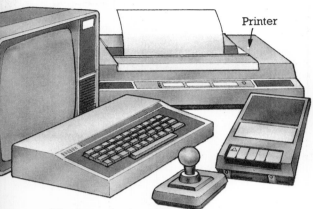

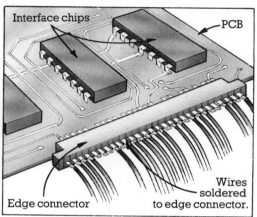

The input device on a microcomputer may be a keyboard, or cassette recorder, or joystick and the output a TV screen or printer.* On other machines the inputs are usually "sensors" which change physical quantities such as heat, speed or the rate of flow of a liquid into electrical signals, and the outputs are most often motors which change electrical signals into movement.

Input and output equipment is connected to the microprocessor by wires usually soldered to long, thin plugs called edge connectors which clip onto the edge of the printed circuit board. Many input and output devices work with signals which are different from binary signals. Special chips called interface chips are used to convert the electrical signals from inputs and outputs to binary and back.

Finding input and output equipment

A microprocessor has no way of knowing what input or output devices are linked to it or where they are. To enable the chip to exchange information with input and output equipment, they are made to look like memory locations holding a byte of data. Each input and output device has its own address and is linked to the address, data and control buses. Information is sent to and from the inputs and outputs in exactly the same way as it is from memory locations.

Some output equipment has two addresses, one to which the microprocessor sends data and a second, called a status address, from which the microprocessor can get information about the output device.

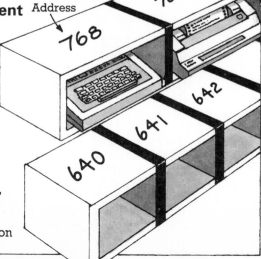

*The input and output devices on a computer are often called peripherals.

Converting signals

Below are some examples of the different kinds of electrical signals produced by input equipment and how these are converted for use by the microprocessor.

Analogue to digital

The electrical signal produced by a computer keyboard is completely different from that produced by, say, a heat sensor. On a keyboard each key creates a separate signal which is either there, when a key is pressed, or not there, when the key is not pressed. These are called digital signals. The signal from a heat sensor is always there, but constantly varying as the temperature goes up or down. It is called an analogue signal (also spelt analog).

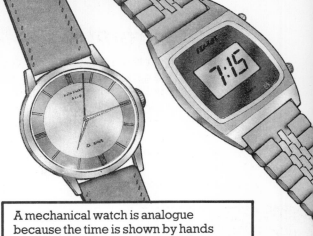

A mechanical watch is analogue because the time is shown by hands which move constantly round the face. An electronic watch is digital because it shows the time as a series of separate little jumps.

A microprocessor can only deal with digital signals, which are either 0 or 1 with no in-between states. Before analogue information, such as temperature or speed, can be sent to a microprocessor it has to be converted to a digital form. This is done by passing the analogue signal through an interface chip called an analogue to digital converter (ADC, for short). The ADC measures the analogue signal at regular intervals (e.g. a thousand times a second), producing a series of values which are coded into binary and sent to the microprocessor.

Serial to parallel

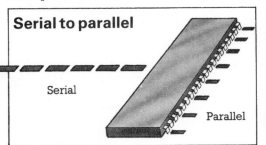

Electrical signals representing data move round a microprocessor eight at a time, each one on a separate track. These are said to be "in parallel". In some equipment however, the electrical signals travel one after the other along a single track. These are "in serial". Cassette tapes which hold computer programs store the bytes of data in serial. Before going to the microprocessor the programs pass through an interface chip, which sends the serial signals on in parallel. The chip is called a serial to parallel converter.

Fast to slow

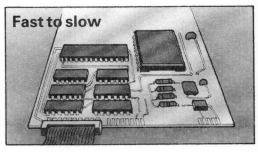

The output equipment linked to a microprocessor often works much more slowly than it does. For instance, a microprocessor might send a thousand characters* to a printer in less than a second, but it takes the printer a minute or two to print them out. To allow for this certain areas of the RAM chips, called buffers, are used to store the character codes while they are waiting to be printed.

*A character is a letter, number or symbol.

More about input and output

The pictures on these two pages show some of the ways in which information can be translated into electrical signals and electrical signals changed back into something which is useful in the real world.

A keyboard taken apart.

Encoder chip

Group of wires called a ribbon cable.

PCB

Electrically sensitive pads

Computer keyboard

Some computer keyboards have an electrically sensitive pad behind each key. The pads are laid on a sheet of plastic and linked with electrical tracks into a grid of rows and columns. When a key is pressed signals for its row and column are sent to an interface chip called an encoder. This recognizes which key has been pressed and produces the binary code for it.

The keyboard and its interface chip have a single address and when the microprocessor fetches data from the keyboard address it is in fact taking it from the interface chip which has produced the binary code. When the microprocessor in a computer is ready to receive the user's instructions it fetches bytes from the keyboard address one after the other and puts them in the RAM chips.

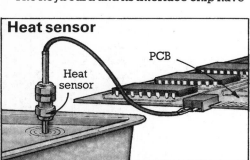

Heat sensor

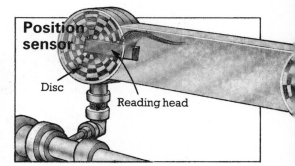

Position sensor

The sensor shown in this picture converts the temperature of a liquid (e.g. the water in a washing machine) to an electrical signal. The sensor contains a small electronic component called a thermistor which allows different amounts of electric current to flow through it according to how hot it is. In this way a varying temperature is converted to a varying electric current. This current is analogue and must be converted to digital before it can be used by the microprocessor (see page 31).

Each joint of a robot arm is fitted with a sensor called an optical position encoder, which lets the microprocessor know where the arm is. The encoder has two parts – a flat disc attached to the moving part of the robot and a "reading head" attached to the stationary part. The flat disc is divided into segments, each one with a different pattern of black and white. As the arm moves, the reading head produces a set of binary signals which match the black and white pattern of the segment beneath it. In this way binary position codes are produced.

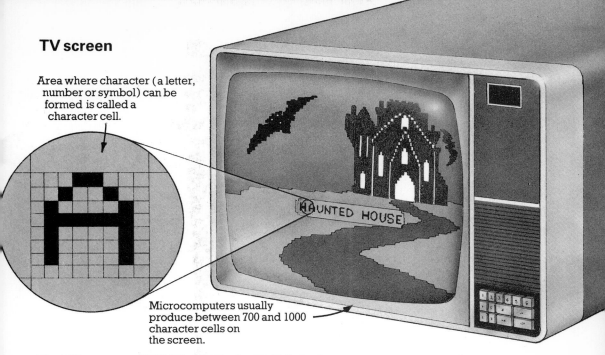

TV screen

Area where character (a letter, number or symbol) can be formed is called a character cell.

Microcomputers usually produce between 700 and 1000 character cells on the screen.

The TV screen or VDU (Visual Display Unit) linked to a computer converts the microprocessor's electrical signals into patterns of light and dark to form words, numbers and pictures which humans can understand.

Each area of a TV screen where a character can be formed has a separate address. When the microprocessor has information to display it sends a binary code for each character to an interface chip called a character generator.* The character generator translates the code into a set of signals which will light up a pattern of dots at one address on the screen to form the character specified by the code.

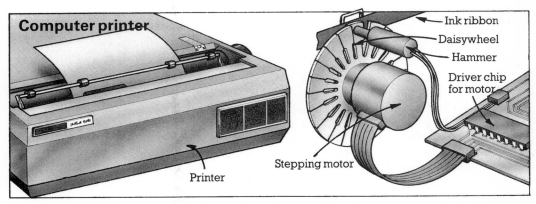

Computer printer

A small "stepping motor" is used to translate electrical signals to movement in a computer's printer. A stepping motor moves round a small, fixed amount each time it receives an electrical signal. The motor turns a plastic disc called a daisywheel which has numbers, letters and symbols at the edge.

When the microprocessor in a computer has information to be printed out it sends binary codes for letters and numbers to the printer's address. Chips inside the printer itself convert the codes into the number of signals necessary to bring the required letter under the hammer.

Many printers have two addresses. The second is called a status address and it holds data about the printer, such as whether it is ready to receive information, or whether it is busy printing, or has run out of paper.

*The character generator is usually an extra ROM chip. On some small computers, though, the character generator circuits are on the main ROM chip.

Inside the ALU

All the data fed into a microprocessor is eventually worked on inside the Arithmetic Logic Unit. This is the part of the microprocessor which performs calculations and makes decisions. It does this by sending the bytes of binary code through circuits called logic gates. These gates are the key to how a microprocessor works.

There are several different types of logic gates made by arranging the transistor switches in different ways. The gates are designed to produce different signals according to what signals they receive, as shown below.

How logic gates work

The tracks which carry signals into a logic gate are called inputs and the track which carries the signal out is called an output. Most logic gates have two inputs and one output and they are organized in groups to deal with the eight-bit binary codes of data.

There are four main types of logic gate in the ALU and you can see how they work in the pictures below.

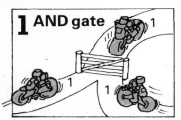

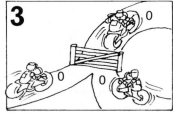

These three pictures show how an AND gate works. It has two inputs and one output. It will send out a binary 1 signal if it receives one on both its inputs, as shown in the first picture. If it receives only one 1 signal or none at all, it sends a binary 0 signal (2nd and 3rd pictures).

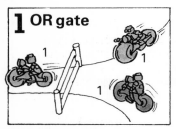
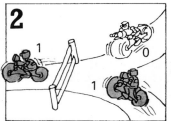

An OR gate will send a binary 1 signal if it receives one on either or both of its inputs.

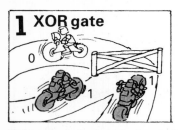

XOR stands for "exclusive OR" gate and this gate is like an OR gate except that it only sends out a 1 signal if it receives a 1 signal on one *or* the other of its inputs, but not both.

A NOT gate has only one input and it always sends the opposite of what it receives. So if its input is a 0 its output is a 1 and vice versa.*

---

*Another name for a NOT gate is an inverter because it inverts the signal it receives.

Complicated decisions

Each type of logic gate can only make a very simple decision, by recognizing whether it has received a certain piece of information. The ALU in a microprocessor often has to make complicated decisions affected by lots of factors. To do this, gates are arranged in groups as shown below.

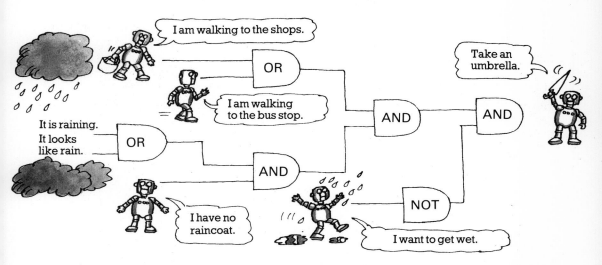

The circuit of gates is designed so that the output of one gate is routed into the input of the next. Each gate is a decision point and the decision made becomes one of the inputs of the next decision point until the final conclusion is reached. As long as a problem can be broken down into a series of logical decision steps it can be worked out by the logic gate circuits in a microprocessor.

A human logic circuit

You can get a group of people to make a model of a logic circuit. Each person is a gate. Their shoulders are the two inputs and one arm is the output. Everyone needs to learn the input and output rules for the gate they are (the rules are given on the opposite page). You also need people to give the initial inputs to the first gates as shown in the pictures below.

If your shoulder is touched the input is a binary 1, if it is not touched the input is a binary 0.

Raising your arm shows a binary 1 output. To pass this on to the next gate, you must touch their shoulder with your hand.

A logic circuit to build

The pictures below show how logic gates could be used to control a piece of equipment such as a green traffic light on a level crossing. You can find out how to build a circuit like this on the next few pages. The electronic circuits which control real level crossings are much more complex but they work in a similar way.

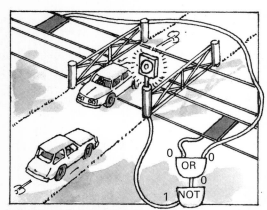

The green light needs to be on most of the time to let cars through, but if a train comes from one or both directions the light must be switched off. A logic circuit to do this can be made by combining an OR with a NOT gate as shown above. The inputs to the OR gate are two switches laid across the railway tracks. These send a binary 1 signal (i.e. a high voltage signal) to the OR gate when a train runs over them. An OR gate sends on a binary 1 if it receives one on either or both of its inputs, so if a train came it would send a 1 to the NOT gate. The NOT gate changes the 1 to a 0 (a low voltage signal) which switches off the green light.

An OR gate combined with a NOT gate is called a NOR gate. You can build a NOR gate using a single transistor and two components called resistors.

NOR gate circuit

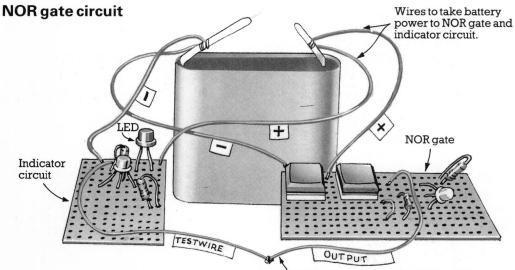

This picture shows what the NOR gate looks like when it is built. To test the gate you need a second circuit called an indicator circuit. This is made using a component called an LED which lights up when it receives a binary 1 signal.*

*You can find out more about all the components on page 44.

Equipment you need

To make the NOR gate you need a battery, a few electronic components shown below and soldering equipment. You can find out how to recognize and use the different components, and how to solder, on pages 44-45. You can buy components from an electronics shop or by mail order through an advert in an electronics magazine.

Things to buy
A 4½ volt battery, about 150cm stranded copper wire, two pieces of 0.1 inch pitch Veroboard, each with 10 tracks and 24 holes.

Components for NOR gate
One BC107 transistor
Two 1K ohm (Ω) resistors
Two push-to-make, 2-pin, PCB-mountable switches suitable for 0.1 pitch Veroboard

Components for indicator
One green LED
One BC107 transistor
One 1K ohm resistor
One 270 ohm resistor

1. Building the NOR gate

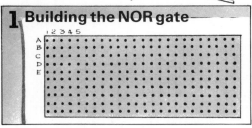

Lay one of the Veroboards on a sheet of paper with the tracks horizontal at the back and make a grid as shown above, giving the tracks letters and numbering the holes.

2.

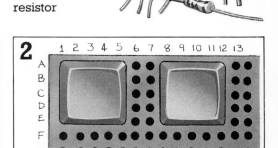

First solder the legs of one of the switches into holes A4 and E2, and the legs of the other into A11 and E9.

3.

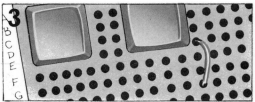

Cut a piece of wire about 1½cm long and strip off a few millimetres of plastic at each end. Solder the wire into holes E13 and H13.

4.

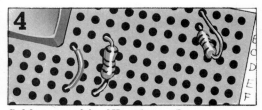

Solder one of the 1K resistors (brown, black, red stripes) into holes D15 and H15 and the other into A21 and B21.

5.

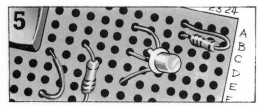

Position the transistor with the collector leg in B18, the base in D18 and the emitter in F18. (You can find out which leg is which on page 44.)

6.

Strip the plastic from the ends of two 30cm wires and one 6cm wire. Solder a 30cm wire into A6 and label it "+". Solder the other 30cm wire into F1 and label it "−". Solder the 6cm wire into B14 and label it "Output".

Building the indicator circuit

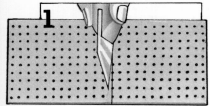

1. Break the other Veroboard so you have a piece with 12 holes and 10 tracks. To break the board score it first with a sharp knife.

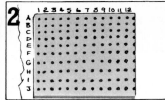

2. Lay the board on a piece of paper with the tracks horizontal at the back and make a grid as shown above.

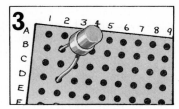

3. Solder the LED into holes B2 and D2. The leg next to the flat edge (or the thicker leg, see page 44) must go in hole D2.

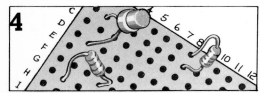

4. Solder the 1K resistor (brown, black, red stripes) into G2 and F4. Solder the 270 ohm resistor (red, purple, brown stripes) into A8 and B8.

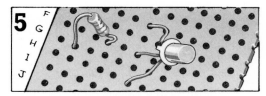

5. Position the transistor with the collector in D6, the base in F6 and the emitter in H6. (See page 44.)

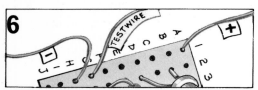

6. Solder a piece of wire 30cm long into hole A1 and label it "+". Solder another 30cm wire into H1 and label it "−". Then solder a short, 6cm wire into G1 and label it "Test wire".

7. To check the indicator works connect it to the battery and touch the test wire against the positive (+) battery terminal. The LED should light up showing there is a high voltage (binary 1) present.*

If you touch the test wire against the negative terminal the LED will not light.

Testing the NOR gate

To test the NOR gate you need to connect it to the indicator circuit. Twist together the output wire of the NOR gate and the test wire of the indicator. You can solder them to make a firmer joint.

Then twist together the + wire of the NOR gate and the + wire of the indicator and hook them onto the positive battery terminal. Do the same with the two "−" wires and hook them to the negative terminal.

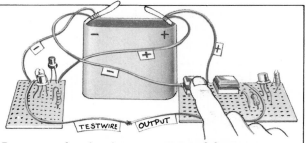

As soon as the circuits are connected the green LED should light up showing that neither of the switches is being pressed (a NOR gate only gives a high voltage output when it gets two low voltage inputs). If you press one or both of the switches the light will go out showing the input has changed to 1 and the output to 0.

*If the indicator does not work check all the components are in the right holes and the joints firmly soldered.

More about logic gates

Logic circuits are useful because they always produce the same output for the same set of inputs. A logic gate with two inputs, e.g. a NOR gate, can receive four different patterns of binary signals: 0,0; 0,1; 1,0; 1,1. Engineers and chip designers write input and output tables showing all the possible results a logic gate can produce. The table for a NOR gate is shown below.

INPUT A	INPUT B	OUTPUT
0	0	1
1	0	0
0	1	0
1	1	0

Truth table for a NOR gate.

If you have built the NOR gate on the last few pages you can check its truth table. Label the switches A and B and work through the table line by line pressing switch A every time input A reads 1 and switch B each time input B reads 1. The green LED shows what the output is.

The proper name for an input and output table is a truth table. If you like you can look back at the descriptions of AND, OR and XOR gates and try to write a truth table for each of them. You can check your answers on page 47.

The story of logic gates

The principles behind electronic logic gates and truth tables were first thought of by an Irish mathematician called George Boole in about 1850. He was studying the way humans make decisions and probably never dreamed his ideas would be used to design electronic circuits.

In his efforts to work out how humans make logical decisions he realised there are three basic thought processes involved, which he called AND, OR and NOT. These are the thought processes which logic gates can imitate.

Boole invented truth tables to show that, provided you could rely on a statement to be true or false (and nothing in between), it was possible to predict all the outcomes of a particular thought process. Boole's truth table for an AND thought process is shown below. He also wrote truth tables for OR and NOT thought processes.

STATEMENT A	STATEMENT B	OUTCOME
e.g. The sun is shining. 	e.g. I am wearing a fur coat. 	e.g. I am too hot. 
FALSE	FALSE	FALSE
TRUE	FALSE	FALSE
FALSE	TRUE	FALSE
TRUE	TRUE	TRUE

If you change the word true to binary 1 and false to binary 0, this truth table is exactly the same as the one for an electronic AND gate. You can check this by looking at the AND gate truth table on page 47.

It was when people realized they could build electronic circuits which would produce the same results as logical human thought processes, that microprocessors and computers became possible.

How the ALU does arithmetic

By combining logic gates in a particular pattern it is possible to make a circuit which will add two numbers. To understand the circuit you need to know the rules for adding in binary and these are shown below.

The adding circuit can only add one binary digit to another. To add eight-bit binary codes the ALU needs a set of eight adding circuits linked together. This group of circuits is called an adder.

How to add in binary
The rules for adding in binary are in fact the same as for adding in decimal.

In decimal the largest digit is 9. When the result of adding two digits is more than 9 (e.g. 5+5 or 9+7) the answer is shown by carrying 1 and starting again at 0 in the right-hand column.

In binary the rules for adding are the same. However, since the largest digit in binary is 1, as soon as you add 1+1 you have to carry. Look at the examples above.

Converting binary to decimal

BINARY	1	1	1	1
DECIMAL	8	4	2	1

$0100 = 4$

$0101 = 4+1 = 5$

$1001 = 8+1 = 9$

Here you can see how two four-bit binary numbers would be added. You can check the addition by converting all the numbers to decimal, as shown on the right.

In a binary number each digit has a value twice as big as the one on its right. To find the decimal equivalent of binary numbers you need to double the decimal number each time you move one place to the left in the binary number, as shown in the chart above. Using the chart you can work out that binary 0100 is 4 in decimal, 0101 is 5 and 1001 is 9 (4+5=9).

Designing an adding circuit

To design an electronic circuit which will add two binary digits, you would need to make a truth table showing all the possible sums and answers two digits could produce, and then think of a circuit which has the same truth table.

Since the binary number system has only two digits, only four different additions are possible.* These are shown below. To write them as a truth table, you make the digits to be added the inputs and the answer the output. In fact, the table needs two outputs, one for the result part of the answer and one for the carry part.

Possible sums	INPUT FOR FIRST BINARY DIGIT	INPUT FOR SECOND BINARY DIGIT	OUTPUT FOR RESULT	OUTPUT FOR CARRY
1. 0 + 0 = 0	0	0	0	0
2. 1 + 0 = 1	1	0	1	0
3. 0 + 1 = 1	0	1	1	0
4. 1 + 1 = 10	1	1	0	1

Which logic gates to use

A circuit to produce the truth table for adding (shown above) can be made using just two logic gates. An XOR gate will produce the output for the result and an AND gate will produce the output for the carry. (You can check this by looking at the AND and XOR truth tables on page 47.) The inputs, that is, the digits to be added, have to be fed to both gates at once as shown below.

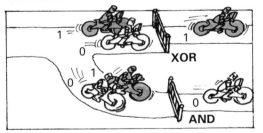
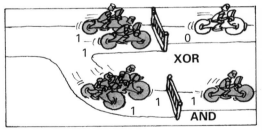

These pictures show how the adding circuit would add 1+0 and 1+1. Inside the ALU there is a set of eight circuits like this, each one to add each of the bits in a byte of binary code. There is also a second set of circuits which add the carry from one adding circuit into the adding circuit on its left. If the last bit of a byte produces a carry, this is indicated by setting a carry flag in the Flag Register (see page 25).

How other arithmetic is done

The ALU does every kind of arithmetic by adding in its adding circuit. To subtract numbers, the ALU makes one of them negative and adds. Multiplication is done by adding over and over again and division is done by subtracting over and over again. Even complex calculations, such as finding square roots, are broken down into simple steps each of which involves only adding. A set of rules for doing complex calculations is called an algorithm. In a pocket calculator chip the ROM circuits hold an algorithm for every mathematical operation the calculator can do.

*In the decimal number system, which has ten different digits, adding two digits produces a hundred possible sums.

Story of the chip

The origins of the chip lie in the research and development of electronics (the control of electricity), which progressed at an amazing speed during the first half of this century. Electronic components were first used to reproduce sound in radios, and to regulate the speed of an electric motor. On these two pages you can find out about the development of the very first electronic components and how the chip became possible.

It was only at the beginning of this century that scientists learned they could control an electric current. Electronics began in 1904 with the invention of the thermionic diode valve, or diode, by Ambrose Fleming. The diode allowed a current to flow through it in one direction only. It worked by emitting electrons (these are electrically charged particles) from a heated wire inside a vacuum-filled glass tube. The electrons were attracted to a metal plate at the other end of the tube, and so the current could flow through in that direction.

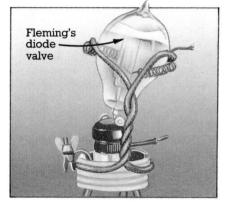

Fleming's diode valve

Later that year a man called Lee de Forest found that by putting a grid of metal in the middle of a diode he could control and vary the size of the current passing through it. This invention, called the triode valve, was very important to the development of radio and television because it allowed currents to be amplified (made larger). It was also discovered that the triode valve could act as a switch. This led to the first electronic computers in the 1940s: Colossus in England and ENIAC in the USA. ENIAC stands for Electronic Numerical Integrator and Calculator. The machine had an incredible 18,000 valves. It filled a huge room and generated an enormous amount of heat. Engineers had to work nearly 24 hours a day replacing valves which overheated.

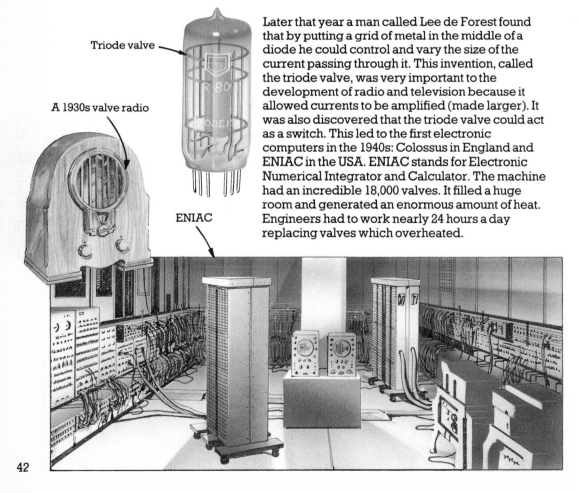

Triode valve

A 1930s valve radio

ENIAC

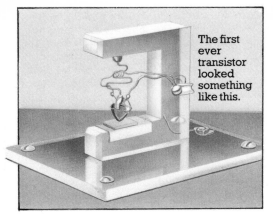

The first ever transistor looked something like this.

In 1947 three American scientists, Bardeen, Brattain and Shockley, invented a device which became known as a transistor. The transistor could do the same as a triode valve but was small, solid and did not need heating. Originally transistors were made from a semi-conductor called germanium. Later silicon was used.

Transistors were first used commercially as amplifiers in hearing aids, but gradually replaced valves in televisions, tape recorders, radios and much larger pieces of equipment such as telephone exchanges. They were first used in computers about 1953.

In 1958 an American called Jack Kilby working at Texas Instruments had the idea of putting two transistors onto one crystal of silicon and made the first integrated circuit. The need for miniaturization created by the American space and defence programmes led scientists to find new ways of squeezing more and more components onto a single chip 5mm square. The pictures on the right give some idea of the speed of progress.

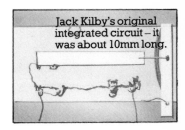

Jack Kilby's original integrated circuit – it was about 10mm long.

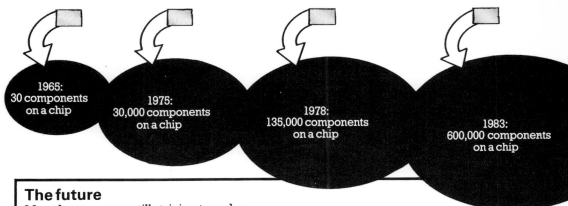

1965: 30 components on a chip

1975: 30,000 components on a chip

1978: 135,000 components on a chip

1983: 600,000 components on a chip

The future

Manufacturers are still striving to make chips more powerful by increasing the number of components on them and the speed at which they work. At present research is being done into a very fast, low power switch which would replace the transistors on chips. The new super switch is called a Josephson junction. It was first thought of in 1962 by a Cambridge University scientist called Brian Josephson and has been developed since then by Bell Laboratories in the USA. A Josephson junction is two thin layers of metal separated by an even thinner insulating layer. At temperatures hundreds of degrees below freezing the metals become "superconducting" and the junctions switch on and off ten times more quickly than the fastest present day transistors. Scientists believe that using Josephson junction technology a computer the size of a baseball could process all the information now handled by a room-sized computer, and do it faster, though it would need to work at sub-zero temperatures.

The next development scientists are talking about is growing chips biologically.

Hints on building circuits

These two pages give information about electronic components and soldering, which will help you build the logic circuit described on pages 36-38.

Transistors

Bipolar and MOSFET transistors look exactly the same from the outside.

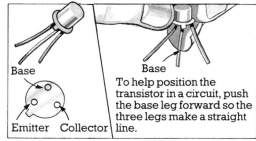

Base. Emitter. Collector.

Base. To help position the transistor in a circuit, push the base leg forward so the three legs make a straight line.

The transistors you need to build the logic circuit are called bipolar, or junction transistors. They are slightly different from the ones described on page 8 (which are called FETs, short for Field Effect Transistors), although they switch a current on and off in the same way.

The three legs of a bipolar transistor are called the emitter, the base and the collector. A current will only flow through the transistor between the emitter and the collector if a voltage is sent to the base leg. The picture above shows how to recognize which leg is which on the BC107 transistor.

Resistors

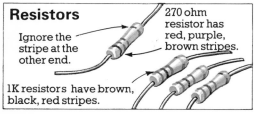

Ignore the stripe at the other end.

270 ohm resistor has red, purple, brown stripes.

1K resistors have brown, black, red stripes.

Resistors reduce the current from the battery to a level the other components in the circuit can cope with. Their strength (that is, the amount by which they reduce the current) is measured in ohms* and shown by the group of three coloured stripes at one end of the resistor. When talking about components, K stands for 1000, so a 1K resistor has a strength of 1000 ohms.

LEDs

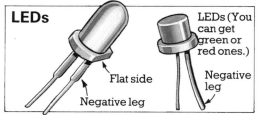

Flat side. Negative leg.

LEDs (You can get green or red ones.) Negative leg.

LED stands for Light Emitting Diode. A diode is a component through which current can only flow one way and LEDs are diodes which glow when current flows through them. LEDs have a negative and a positive leg and it is important to position them correctly in circuits. On some LEDs you can recognize the negative leg because it is thicker. On others the LED's body has a flat side and the negative leg is closest to it.

Using Veroboard

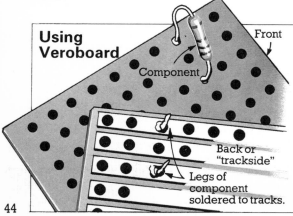

Front. Component. Back or "trackside". Legs of component soldered to tracks.

Veroboard is specially designed for building electronic circuits. It has rows of holes, with strips of copper at the back linking the holes. You fit the legs of the components and the wires from a battery, through the holes and solder them to the copper at the back. Then the electric current can flow to the components along the copper tracks. The size of Veroboard is given as the number of tracks by the number of holes in each track (e.g. 10 tracks × 24 holes).

*The symbol for ohm is Ω.

How to solder

What you need:
- A small soldering iron,
- some cored solder,
- a damp sponge,
- miniature pliers,
- wire cutters or scissors.

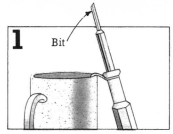

1 Plug the soldering iron in. While it is heating up, support it so that the bit is not touching anything.

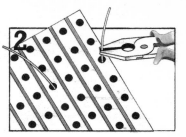

2 To solder a component on to Veroboard, find the right holes and push the legs through. Bend them out slightly with pliers.

3 Touch the end of the solder with the hot bit, so that a drop of solder melts and clings to the bit.

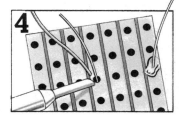

4 Then touch the leg of the component with the bit and the tip of the solder for a second until a drop of solder joins it to the track.

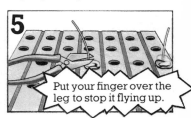

5 Put your finger over the leg to stop it flying up.

Let the joint cool for a few seconds. Then, tilt the board away from you and trim the legs close to the soldered joint with wire cutters.

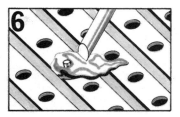

6 It is very important to remove any solder that runs into the grooves between the tracks. Run the hot bit along the track a few times.

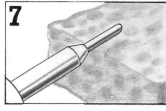

7 After each joint, clean the bit on the damp sponge and *remember to unplug the soldering iron when you have finished.*

Desoldering

To remove, or desolder, a component, wedge the tip of a pencil between the component's legs on the top side of the Veroboard. Ask someone to tilt the Veroboard and hold the pencil, easing the component out as you melt the joints on the tracks with the soldering iron.

How to tin wire

If you are using stranded wire, it helps to coat (tin) the ends with a layer of solder, making it easier to push them through holes on the board.

Strip about a centimetre of the plastic coating off the wires using wire strippers. Twist the strands of wire together.

Put something heavy on the wire to hold it still. Stroke the twisted strands of wire with the solder and the bit a few times to coat them.

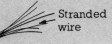
Stranded wire

Be careful – the soldering iron gets extremely hot.

Microprocessor pin chart

The pins of a microprocessor's case carry power and address, data and control signals into and out of the chip. The picture on the right is a diagram, or pin chart of one of the most common microprocessors, the Z80. (Z stands for Zilog, the manufacturer's name.) The pins are numbered 1 to 40, with pin number 1 to the left of the notch at the top of the case. The labels on the pins indicate the signal they carry. Below you can find out what the labels stand for and what each signal does. Many of the labels have a line across the top. This is called an overbar and indicates the pin is active when low. You can find out what this means on page 27.

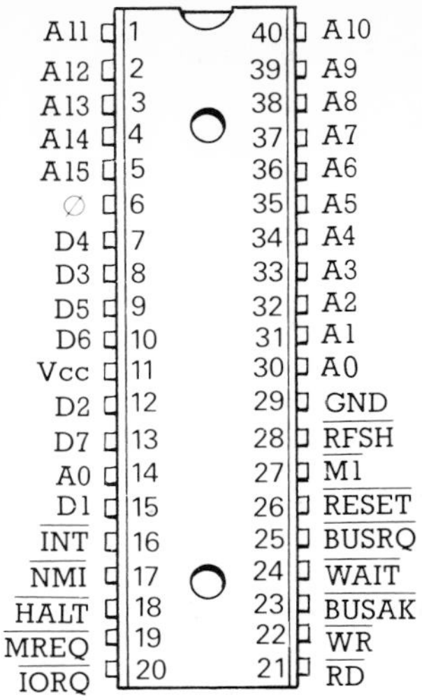

```
A11  [ 1      40 ] A10
A12  [ 2      39 ] A9
A13  [ 3      38 ] A8
A14  [ 4      37 ] A7
A15  [ 5      36 ] A6
 Ø   [ 6      35 ] A5
 D4  [ 7      34 ] A4
 D3  [ 8      33 ] A3
 D5  [ 9      32 ] A2
 D6  [ 10     31 ] A1
 Vcc [ 11     30 ] A0
 D2  [ 12     29 ] GND
 D7  [ 13     28 ] RFSH
 A0  [ 14     27 ] M1
 D1  [ 15     26 ] RESET
 INT [ 16     25 ] BUSRQ
 NMI [ 17     24 ] WAIT
 HALT[ 18     23 ] BUSAK
 MREQ[ 19     22 ] WR
 IORQ[ 20     21 ] RD
```

A0-A15 These pins carry the 16-bit address codes out of the microprocessor. A stands for address and the number refers to the position of the bit in the code.

D0-D7 These pins carry the eight bit data codes into and out of the microprocessor. D stands for data.

Ø This pin carries the clock signal (represented by the Greek letter, phi) into the microprocessor.

Vcc This pin is for the power supply. It is connected to +5 volts.

INT Stands for interrupt. This signal interrupts what the microprocessor is doing to make it react to an outside emergency, for instance, the machine overheating.

NMI Stands for non-maskable interrupt. This is a second interrupt signal which overrides the first one. It is called non-maskable because nothing can override, or mask, it.

HALT This is a signal the microprocessor sends to tell other chips it has stopped work temporarily.

MREQ Stands for memory request. It is a control signal which the microprocessor sends to let the memory chips know there is an address on the address bus.

IORQ Stands for input/output request. This is similar to the memory request signal except that it lets input and output equipment know there is an address on the address bus.

RD Read. The control signal which indicates data is to be read from a memory location or I/O (input/output) device.

WR Write. The control signal which indicates data is to be written into a memory location or I/O device.

BUSAK This and the BUSRQ signals are used when the microprocessor is sharing the address and data buses with other microprocessors. BUSAK stands for bus acknowledge and it is a signal which the microprocessor sends to let another processor know the address and data buses are free for it to use.

WAIT This is a signal sent to the microprocessor to make it wait a few extra clock cycles for something outside to happen. For example, some kinds of ROM chips work more slowly than most memory chips, so the microprocessor must wait longer for data to be fetched from this type of chip. The microprocessor starts waiting when the voltage on the pin goes low and waits until it goes high again.

BUSRQ Stands for bus request. This is the signal sent by another microprocessor sharing the buses, to let this microprocessor know it wants to use them.

RESET This pin carries the reset signal which sets the Program Counter back to 0 when the microprocessor is first switched on (see page 29).

M1 This is a signal the microprocessor sends to other chips to let them know it is fetching an instruction from memory.

RFSH Stands for memory refresh. This is a signal the microprocessor sends to dynamic RAM chips (see opposite) to keep the data they hold intact.

GND Stands for ground. This pin carries the power supply out of the microprocessor. It is called ground because it is connected to 0 volts.

Chip words

ACIA A common interface chip which handles serial/parallel conversions. ACIA stands for Asynchronous Communication Interface Adaptor.

CMOS A type of chip which contains both n- and p-channel MOSFETs (see below). It runs on very little power. CMOS stands for Complementary Metal Oxide Semiconductor.

Dynamic RAM chip A RAM chip which needs to be constantly refreshed with electrical signals in order to retain its data.

EAPROM or EAROM Stands for Electrically Alterable (Programmable) ROM. This is similar to an EPROM chip, described below, but its programs are erased by applying voltages to certain of its pins. Also known as EEPROMs or EEROMs (Electrically Erasable).

EPROM A ROM chip similar to a PROM (see below), but its programs can be erased and new ones written in by a process which involves flooding the chip with ultra-violet light. EPROMs always have a hole in the top of the case for this purpose.

Flip-flop An electrical circuit made using several transistors, which can be in one of two states to represent 0 and 1. Flip-flops are used in memory chips to form memory cells and on a microprocessor to form registers.

LSI Stands for Large Scale Integration. Scale of integration normally refers to the number of components on a chip. In LSI chips there are about 100-10,000.

MOS Stands for Metal Oxide Semiconductor. This describes the technology used to manufacture most chips, which uses metal as the electrical conductor and silicon dioxide as an insulator.

MOSFET Stands for Metal Oxide Semiconductor Field Effect Transistor. There are two types of MOSFETs: n-channel, described on pages 8 and 9, and p-channel which is made with two islands of p-type silicon in a bed of n-type.

MSI Stands for Medium Scale Integration. In MSI chips there are between 10 and 100 components.

NAND gate A type of logic gate which performs the opposite operation to an AND gate. NAND stands for "not AND".

NMOS A type of chip which contains only n-channel MOSFETs (see above). NMOS chips work very quickly.

PMOS A type of chip which contains only p-channel MOSFETs (see above). PMOS chips use quite a lot of power.

PROM A special kind of ROM chip onto which programs can be written after it has been manufactured by melting tiny fuses built into the circuits. PROM stands for Programmable ROM.

SSI This stands for Small Scale Integration. In SSI there are less than 10 components on a chip.

Static RAM chip A type of RAM chip which does not need special signals to enable it to retain the data stored in it while it is switched on. (See also dynamic RAM.)

TTL Stands for Transistor Transistor Logic and describes chips with logic circuits based on bipolar transistors (not MOSFETs).

UART A common interface chip which handles serial/parallel conversions. UART stands for Universal Asynchronous Receiver/Transmitter.

ULA Stands for Uncommitted Logic Array. This is a chip with logic gates which a manufacturer can wire up in different ways according to the use to which the chip will be put by a particular customer.

VIA A type of interface chip which can handle all types of signal conversions (e.g. analogue/digital, serial/parallel). VIA stands for Versatile Interface Adaptor.

VLSI Stands for Very Large Scale Integration. VLSI chips have over 10,000 components.

AND truth table

INPUT 1	INPUT 2	OUTPUT
0	0	0
1	0	0
0	1	0
1	1	1

OR truth table

INPUT 1	INPUT 2	OUTPUT
0	0	0
1	0	1
0	1	1
1	1	1

XOR truth table

INPUT 1	INPUT 2	OUTPUT
0	0	0
1	0	1
0	1	1
1	1	0

Index

Accumulator, 25
ACIA chip, 47
"active when high/low", 27, 46
adding circuit, 41
address, 19, 20, 22, 23, 25, 33
    bus, 19, 20, 29, 30
    input/output, 29, 30
    signals, 46
    status, 30, 31, 33
ALU, 22, 23, 25, 34-35, 40-41
analogue, 31, 32
AND gate, 34, 39, 41, 47
Arithmetic Logic Unit, 22, 23, 34-35, 40-41
Babbage's Difference Engine, 10
Bardeen, 43
base (transistor leg), 44
binary,
    arithmetic, 40-41
    code, 10, 11, 22, 23
    signals, 10, 17, 19, 22, 23, 30, 31
bipolar transistors, 44
bits, 10, 21, 25
Boole, George, 39
Brattain, 43
buffers, 31
bus, 10, 19, 20, 21, 30
    acknowledge signal (BUSAK), 46
    request signal (BUSRQ), 46
bytes, 10, 22, 23, 29
capacitor, 5
carry flag, 25, 41
cassette recorder, 30, 31
Central Processing Unit (CPU), 16
character generator, 33
circuit, definition of, 4
clock, 17, 23, 26, 27, 28-29, 46
CMOS chip, 47
collector (transistor leg), 44
Colossus, 42
components, 5, 42, 44
computer, 6, 18, 29, 30, 39, 42, 43
control,
    bus, 19, 20, 27, 30
    circuits, 22, 23, 26, 28
    lines, 27
    signals, 20, 26-27, 28-29, 46
crash, 23
data, 19
    bus, 19, 20, 21, 30
dedicated microprocessor, 17
digital signals, 31, 32
diode, 42
dopant, 15
drain, on a transistor, 8, 9
dynamic RAM, 46
EAPROM/EEPROM chip, 47
EAROM/EEROM chip, 47
edge connector, 17, 30
eight-bit code, 10, 19
electron beam lithography, 14
electrons, 42
electronic components, 5, 42
electronics, 4, 5, 42

emitter (transistor leg), 44
encoder chip, 32
EPROM chip, 46
Field Effect Transistor, 44, 47
Flag register, 25, 41
Fleming, Ambrose, 42
flip-flop, 47
de Forest, Lee, 42
gate (on a transistor), 8, 9
germanium, 43
Grand Prix racing car, 9
HALT signal, 46
heat sensor, 31, 32
incrementing, 24
input/output,
    address, 29, 30
    devices, 30-33
    request signal (IORQ), 46
Instruction Register, 24, 26, 29
Instruction Set, 26
integrated circuit (IC), 5, 43
interface chips, 17, 30, 31, 32, 47
Interpreter, 18
interrupt signal (INT), 46
inverter, 34
ion implantation, 15
Josephson junction, 43
joystick, 6, 30
junction transistor, 44
k, K, 21, 44
keyboard, 6, 30, 31, 32
Kilby, Jack, 43
kilobyte, 21
LED (Light Emitting Diode), 36, 44
logic gates, 34-35, 39
machine code instruction, 18, 26, 28
megahertz, 28
memory,
    chips, 16, 18, 19, 20, 21, 23, 47
    location, 19, 20, 21
    refresh signal (RFSH), 46
    request signal (MREQ), 46
micron, 14
microprocessor, 4, 16, 17, 18, 19, 20, 22-23, 28, 29
Microprogram ROM, 26
Monitor, 18
MOS chip, 47
MOSFET, 47
NAND gate, 47
nanosecond, 12, 21
NMOS chip, 47
non-maskable interrupt signal (NMI), 46
NOR gate, 36, 39
NOT gate, 34, 36, 39
n-type silicon, 8, 15, 47
ohms, 44
Operating System, 18
optical position encoder, 32
OR gate, 34, 36, 39, 47
overbar, 46
parallel signals, 31, 47
photomask, 14

photoresist, 14
pins, 5, 20, 46
PMOS chip, 47
printed circuit board (PCB), 17, 30, 32
printer, 30, 31, 33
programs, 22
Program counter, 22, 24, 29, 46
PROM chip, 47
p-type silicon, 8, 15, 47
quartz crystal, 28
Random Access Memory (RAM), 16
    chip, 18, 21, 25, 29, 31, 47
Read Only Memory (ROM), 16
    chip, 16, 17, 18, 29, 47
Read/Write signals, 19, 20, 46
registers, 22, 23, 24-25, 26
reset signal, 29, 46
resistor, 5, 44
robot, 7, 22, 30, 32
scale of integration, 47
semiconductor, 8, 43
sense amplifiers, 20, 21
sensors, 30, 31, 32
serial signals, 31, 47
Shockley, 43
silicon, 5, 8, 14, 43
    dioxide, 14
simulation, 13
sixteen-bit code, 19, 46
solder, how to, 45
source (on a transistor), 8, 9
space, 6, 7, 43
square wave, 28
static RAM, 47
status address, 30, 33
stepping motor, 33
thermionic diode valve, 42
thermistor, 32
transistor, 5, 8, 9, 34, 43, 44, 47
triode valve, 42-43
truth tables, 39, 41, 47
T-state, 28
TTL (Transistor Transistor Logic), 47
TV screen, 6, 18, 30, 33
UART (Universal Asynchronous Receiver/Transmitter), 47
ULA (Uncommitted Logic Array), 47
valves, 42
VDU (Visual Display Unit), 33
Veroboard, 44
VIA (Versatile Interface Adaptor), 47
voltage, 8, 9, 10, 27
WAIT signal, 46
washing machines, 7, 30
watches, 6, 31
Working registers, 25
XOR gate, 34, 41, 47
Z1 calculating machine, 11
Z80 microprocessor, 46
Zilog, 46
Zuse, 11

Circuit on page 37 designed by Malcolm Smith.
Thanks to CBA Inc/New Scientist for the pictures on page 13, and to The Marconi Company for the picture on page 42.

First published 1983 by Usborne Publishing Ltd, 20 Garrick Street, London WC2E 9BJ, England.
© 1983 Usborne Publishing
All rights reserved. No part of this publication may be reproduced, stored in a retrieval system or tansmitted in any form or by any means, electronic, mechanical, photocopying, recording or otherwise, without the prior permission of the publisher.
The name Usborne and the device are Trade Marks of Usborne Publishing Ltd.

Printed in Spain by Printer Industria Gráfica, S.A. Depósito legal B-29768/83